KGB TRAINING MANUAL

TRADECRAFT AND COVERS

КОНСПИРАЦИЯ В РАЗВЕДЫВАТЕЛЬНОЙ
РАБОТЕ НА ТЕРРИТОРИИ СОВЕТСКОГО
СОЮЗА С ПОЗИЦИИ ВЕДОМСТВ ПРИКРЫТИЯ

1988

Translated from the original Russian
by Major Christoph P. Schwanitz (Ret.)

Conflict Research Group
London, 2025

Original edition published for internal use by the Ministry for State Security (KGB) of the Union of Soviet Socialist Republics.

This English language translation published by Conflict Research Group, London, United Kingdom, 2025

Copyright Notice

About the Conflict Research Group

Conflict Research Group (CRG) is a non-profit think-tank based in the United Kingdom, dedicated to advancing understanding of the art and science of Unconventional Warfare. With a focus on the academic study of guerrilla warfare, revolutionary warfare, asymmetric warfare, Fourth Generation Warfare, Fifth Generation Warfare and political unrest, CRG's work sheds light on the complexities and nuances of modern conflicts. By bringing critical and key works back into print, the organization serves as a vital resource for academics, policymakers, and military professionals seeking in-depth knowledge in these specialized fields.

At the heart of CRG's mission is the belief that a comprehensive understanding of Unconventional Warfare is essential for addressing contemporary security challenges. The group's research and publications delve into historical and contemporary case studies, exploring the strategies, tactics, and implications of irregular warfare. Through this rigorous analysis, CRG contributes to the development of more effective and adaptable strategies for dealing with non-traditional threats.

One of the key aspects of CRG's work is its publishing arm, which is dedicated to bringing into print seminal works on Unconventional Warfare. The group's publications cover a wide range of topics, from historical accounts of guerrilla movements to theoretical analyses of contemporary conflict dynamics and of course reprints of historical official publications. By making these works accessible to a broader audience, CRG aims to enrich the discourse on Unconventional Warfare and contribute to the development of more nuanced and effective approaches to resolving conflicts and disrupting, degrading and defeating unconventional threats.

CRG's research is categorised by its interdisciplinary approach, drawing on insights from military history, political science, sociology, and international relations. This holistic perspective allows the organization to address the multifaceted nature of unconventional warfare, considering not only military tactics, but also the granularity of the political, social, and economic dimensions of conflicts. Through this comprehensive approach, CRG provides a deeper understanding of the root causes and long-term implications of irregular warfare.

Publisher's Note

This English translation of *Tradecraft and Covers* (1988) is based on a Soviet-era KGB manual that was never declassified. The original document, produced by the KGB's First Chief Directorate, was intended for internal use in intelligence training and operational instruction. While the Soviet Union no longer exists, this text serves as an important historical resource on Cold War espionage tradecraft.

The translation presented here is the result of meticulous effort to ensure accuracy, clarity, and readability for an English-speaking audience. It includes refinements, contextual clarity, and structural modifications to make the material accessible without altering the meaning of the original text. Additionally, this edition features a *Translator's Note*, *Editor's Introduction* and an explanatory chapter about the KGB to provide historical and operational context.

Neither the translator or the publisher endorse or promote espionage activities but present this work purely for educational, historical, and academic research purposes.

Translator's Note

I grew up in East Germany during the 1970s and 80s at the height of the Cold War, with an uncle who was a high-ranking official in the notorious and now thankfully defunct East German Ministerium für Staatssicherheit or Ministry for State Security, more commonly known today as the Stasi.

I did not know it at the time, but before moving into his highly prestigious and important role as Stasi chief for East Berlin, my uncle had spent a large part of his career in the Stasi's Hauptverwaltung Aufklärung section (HVA) which dealt with foreign intelligence operations. This was the GDR's equivalent of the Soviet KGB or the American CIA or the British MI6.

As a young man, particularly during that chaotic period immediately following German Reunification, this all conspired to give me a deep interest in the intricacies of intelligence operations and ensured that I would later seek to forge a career in the field of intelligence. Of course, I would go on to do just that, serving with several different operational units within the German Federal Army until my retirement from the Bundeswehr in 2017.

As someone who had spent decades studying Russian language, culture, military and intelligence structures in the course of my work as an officer in the Army, I knew that translating these manuals would require more than just basic linguistic proficiency. It would demand an intimate knowledge of the nuances of each language, as well as a deep understanding of the cultural, historical, technical and operational contexts in which they were written.

One of the more significant challenges I faced, and the one which is least likely to be of any great interest to a reader of this book, was navigating the complexities of Russian grammar and syntax. Unlike German, which is known for its strict rules and conventions, Russian has a more relaxed system that can make it difficult at times to accurately convey meaning. For example, Russian word order often prioritizes grammatical function over semantic content, making it essential to carefully consider the context in which each sentence appears.

Furthermore, Russian relies heavily on prepositions and case endings to convey subtle shades of meaning, whereas German tends to rely more on verb conjugation and adverbial phrases. This meant that I had to be particularly mindful when translating individual words or phrases, as their meanings could shift significantly depending on the surrounding context.

Another significant challenge was capturing some of the more obscure technical jargon used in these manuals. With many of the KGB manuals in the cache dating to the 1960s and 70s, some of the old Soviet terminology has become obsolete and has been replaced by other terms within the Russian Federation intelligence services. The manuals display a wide array of obsolete and specialized terms for various aspects of intelligence operations, from types of agents to counter surveillance techniques to clandestine communications methods.

As someone who is very familiar with German and other NATO partners' intelligence operations and with their jargon and acronyms, I found myself constantly referencing my own knowledge base to ensure that I accurately conveyed the intended meaning of any given passage that included technical operational language. In those few cases where I could not be 100% sure of a technical term's meaning, I simply extrapolated to the best of my ability.

The fact that many of these manuals remain classified in Russia even today speaks volumes about their significance and relevance. It's likely that some are still being used by Russian Federation intelligence services to train new personnel, while others would have no doubt been declared obsolete but remain sensitive due to the nature of their contents.

I have a responsibility to ensure that these manuals are translated accurately and with appropriate sensitivity. It's not just about conveying technical information; it's also about respecting the cultural and the operational context in which they were written. In many ways, translating these KGB tradecraft training manuals was akin to conducting an archaeological excavation into the past. Each sentence or phrase revealed a piece of history that had been hidden away for decades, waiting to be uncovered and shared with the world.

As someone who has spent years studying Russian language and culture as well as evaluating the potential threats which an adversarial Russian Federation might in the future pose to my homeland and our NATO partners, I'm proud to have played a role in making this significant historical material available for public consumption.

I would like to thank "DC" and "CB" from Conflict Research Group for assigning me the delicate but critical task of translating this important material. Having become well informed of the vital work being undertaken by Conflict Research Group, I am honoured to be of service even in this small way.

I would like to thank my beloved wife, Birgitt, for dealing with my many absences and long days spent locked away in my study working on this material and accepting it all with grace and good humour.

I would like to also thank Birgitt for her assistance in helping me translate certain more complex passages from German to English and for proof-reading the final manuscript to correct my abysmal English language grammar. As always, without her, I would be diminished.

Please note that any errors or omissions in these translated pages which may serve to detract from the original Russian language documents are mine and mine alone.

Christoph P. Schwanitz,
Major, KSA (ret.)
Görlitz, 2025

About the KGB

The KGB was the foreign intelligence and domestic security agency of the Soviet Union. It was established on the 13th of March, 1954, soon after the death of Soviet dictator Josef Stalin and it was dissolved with the fall of the Soviet Union on the 3rd of December 1991. The KGB's First Main Directorate was split off and became the Russian Federation's current foreign intelligence service, the SVR.

In addition to its primary responsibilities for foreign intelligence and domestic counterintelligence, during the Soviet era the KGB also had duties such as safeguarding the country's political leadership, overseeing border troops, and carrying out surveillance of the population.

In this book, we are dealing solely with the foreign intelligence aspects of KGB operations, so we shall look at the KGB's foreign intelligence apparatus.

The KGB's First Main Directorate, also known as the First Chief Directorate, was responsible for intelligence operations outside of the Soviet Union.

The directorate was organised into various directorates, including:

Directorate "R" -   Planning and Analysis,
Directorate "S" -   Illegals,
Directorate"T" -   Scientific and Technical Intelligence,
Directorate "K" -   Counter-Intelligence,

Directorate "OT" -	Operational and Technical Services,
Directorate "I" -	Computers Service (would be known as "IT" today),
Directorate "A" -	Active Measures,
Directorate "RT" -	Operations within the USSR

In addition to the administrative directorates listed above, the First Main Directorate had various "Desks" or "Departments" dedicated to operations in various parts of the world or other specialised functions. These were:

1st Department -	North America
2nd Department -	Latin America
3rd Department -	UK, Australia, NZ, Scandinavia, Malta
4th Department -	East Germany, Austria, West Germany
5th Department -	France, Spain, Portugal, Luxembourg, Switzerland, Greece, Italy, Yugoslavia, Albania, Romania
6th Department -	China, Laos, Viet Nam, Cambodia, North Korea, South Korea
7th Department -	Thailand, Indonesia, Malaysia, Singapore, Japan, Philippines
8th Department -	Afghanistan, Turkey, Iran, Israel
9th Department -	English-speaking countries in Africa (South Africa, Rhodesia/ Zimbabwe, Tanzania, Nigeria, etc.)
10th Department -	French-Speaking Countries in Africa
11th Department -	Liaison with other communist countries' intelligence services particularly Cuban and Warsaw Pact nations (was previously known as the "Advisor's Department")
12th Department -	Covers
13th Department -	Covert Communications
14th Department -	Forgeries
15th Department -	Operational files and archives

16th Department -	Signals intelligence
17th Department -	India, Pakistan, Bangladesh, Sri Lanka, Burma, Nepal
18th Department -	Egypt, Syria, Libya, Iraq, Oman, Saudi Arabia, Kuwait, Sudan, Jordan, Morocco, United Arab Emirates/Trucial States
19th Department -	Soviet Expatriates and Emigres
20th Department -	Liaison with 3rd World / newly independent states

It is the KGB's First Main Directorate which was the publisher of these manuals and is most likely that they were produced by staff of the First Main Directorate's *Directorate OT*, which was responsible for Operational and Technical Support functions.

KGB foreign intelligence networks were operated by a KGB Residence or *Station* as it is known in phonetic Russian. Please note that in these translations, we sometimes refer to the Residencies using the equivalent CIA term "Station". This is simply to reduce the possibility of confusion and to differentiate between a Residence and a private residence such as those used as safe houses or clandestine postal addresses. Similarly, within the translation in these situations, we will refer to the KGB Residence's "Resident" using the CIA term "Station Chief".

The KGB Resident or Station Chief was a legal intelligence officer usually operating under diplomatic cover as a "cultural attache" or similar. Diplomatic credentials gave the Resident diplomatic immunity meaning the security forces of the country in which he was operating could never arrest a KGB Resident. At best they could have him expelled from the country like any other diplomat, but this usually had serious diplomatic consequences. Instead, most countries usually worked out fairly quickly who was KGB within their local Soviet embassy and they usually allowed the KGB Resident to operate, but placed him and other Soviet embassy staff under heavy counterintelligence surveillance.

Typically, a KGB Residency was organised into different sections or "lines". Each section had a separate function which supported operations conducted out of any given Residency. These sections could be further categorised into separate functions - Operational and Support.

Operational sections of a KGB Residency were as follows:

Section "EM" - Intelligence and surveillance of the activities of Soviet Emigres in the host country

Section "KR" - Counterintelligence and protective security of the Residency

Section "N" - Support to "illegal" Intelligence Officers in the host country

Section "PR" - Economic, military, political intelligence on the host country or region as well as active measures such as black propaganda

Section "SK" - Surveillance and reporting on Soviet diplomatic staff in the host nation.

Section "X" - Technical intelligence and advanced technology acquisition and transfer.

Support sections of a KGB Residency were as follows:

Section "OT" - Technical support
Section "RP" - Signals intelligence
Section "I" - Information technology

Support staff not assigned to their own specific section included drivers, signals operators, cipher clerks, administrative staff, finance personnel.

Table of Contents

PART I:
TRADECRAFT SECURITY - A CRITICAL PRINCIPLE AND PREREQUISITE FOR THE SUCCESS OF INTELLIGENCE OPERATIONS

PART II:
MAINTAINING TRADECRAFT SECURITY IN THE PROCESS OF ESTABLISHING COVERS AND THEIR OPERATIONAL USE

PART III:
FEATURES OF THE LEGEND DEVELOPMENT, DISGUISE AND DECEPTION TECHNIQUES IN THE PROCESS OF AGENT-OPERATIONAL ACTIVITY

Editor's Introduction

The original Russian language manual this English translation is based on was found on the deep web in a cache of scanned older Soviet KGB training materials in a folder on a Russian language .onion site. It is believed that these materials were posted by a dissident many years ago, perhaps even as long ago as 2010 or 2012 based on file metadata. The cache was later posted on the surface web, where to this day scans of the original Russian language documents can still be found through a simple search on any search engine.

Various think-tanks from English-speaking countries had made promises to translate and publish these materials, but despite waiting over five years for them to do so, no apparent progress has been made. With the Russian invasion of Ukraine in February 2022, it appears that translation and publication of the KGB training manuals is no longer a priority for these organizations. As a consequence, and with no clear end to the Ukraine War in sight at the time of writing, we have gone ahead and translated and published the KGB manuals from the cache ourselves.

Please note that we are not the first to publish English language translations of some of these materials. Circa 2020, enterprising persons unknown, in a blatant cash-grab, ran a couple of these documents through some translation software, probably Google, before dumping the resulting unedited text into a book format for publishing on Amazon. We purchased a copy of each of these translations to see whether there would still be a requirement for our professionally translated editions. Sadly, all were largely unreadable, therefore, we pushed ahead with our project.

As Conflict Research Group deals mostly with unconventional warfare, resistance, and inform/influence operations from the perspective of non-state actors, it would seem to the casual reader that espionage training materials from a former nation-state intelligence agency such as the Soviet KGB would fall well outside our remit.

This is simply not the case. During the Cold War, the Soviet Union, its Warsaw Pact satellites and other communist states such as the People's Republic of China and the Democratic People's Republic of Korea invested many billions of dollars in to supporting subversive and revolutionary groups fighting against western interests from Southeast Asia to the Middle east, to Latin America, to Southern Africa. Soviet support for such groups was not limited to weaponry and war materiel, but also included training in communist political theory, revolutionary and guerrilla warfare and of course, in clandestine tradecraft to allow members of a revolutionary or terrorist group to organize, plan and conduct their activities in secret.

Western-trained security forces typically used extremely effective British, French or American counter-intelligence and counter-insurgency methods to detect and destroy insurgent undergrounds or espionage rings at or before their nascent stage, so there was a requirement for guerrilla or terrorist groups sponsored by the Soviet Union to be given the most effective tradecraft training available in the communist world, and that came from the KGB's First Main Directorate.

Two English language resources which closely follow KGB procedures and concepts can be found in the 1980s-era South African Communist Party pamphlet *How to Master Secret Work* and in the 1970s-era document *Security and the Cadre* produced by a Puerto Rican separatist group operating in the US, the Fuerzas Armadas de Liberacion Nacional (FALN). Anyone reading through those two sources and then reading one of these KGB manuals will soon find examples which appear in all three, sometimes almost word-for-word.

Unlike some Western intelligence services such as the CIA, which train personnel in very specific, complex tradecraft techniques and methodology (some involving literal magician's sleight of hand), the KGB instead concentrated on teaching its personnel general concepts. This forced the KGB operative to become highly adaptable and imaginative in putting those concepts into action in the field. This lack of a "toolkit" of relied-upon tactics, techniques and procedures meant no clear patterns were set, making it just that much harder for western counterintelligence services to anticipate the specificities of a KGB intelligence officer or agent's tradecraft in the field.

In closing, I would like to thank Major Chris Schwanitz for his accurate translations of these materials, as well as for standing by for almost 18 months while we decided whether or not to go ahead with this project. I would also like to thank "DC" and her OSINT team for backtracking the circumstances of how the original scanned documents came to be posted online. Finally, I would like to thank you, the reader, for your interest in this project and for your support of CRG by purchasing this book.

CB
London, 2025

Annotation

Tradecraft in Intelligence Work on the Territory of the Soviet Union from the Perspective of Cover Agencies - Analytical Overview

This publication consolidates the accumulated experience of officers from Directorate "RT" of the First Chief Directorate (PGU) in maintaining tradecraft security through the use of cover institutions in the execution of intelligence tasks within the territory of the Soviet Union.

It presents a set of recommendations aimed at enhancing the clandestinity of intelligence activities conducted by Directorate "RT" personnel, whether operating within the central apparatus or embedded inside cover agencies.

Introduction

In recent years, the operational environment in most target countries has sharply deteriorated. The intelligence services of the United States and other capitalist states have advanced their methods and means of conducting operations against Soviet intelligence. They are making increased efforts to penetrate the intelligence organs of the USSR and its agent networks. Under these conditions, the successful accomplishment of the tasks assigned to the First Chief Directorate (PGU) depends to a great extent on raising political vigilance, strictly adhering to tradecraft principles, and maintaining strict operational security.

Tradecraft security in foreign intelligence occupies a special place. It is one of the foundational principles—an indispensable element of all Chekist activity. It is the key method of ensuring safety, protecting personnel, and preventing enemy penetration into the cadres and agent networks of the KGB. In Order No. 0100 issued by the Chairman of the Committee for State Security (KGB) of the USSR on 19 February 1986, "On Measures to Enhance Tradecraft Security in the Activities of State Security Organs under Modern Conditions," it was emphasized that "every operational officer must understand that tradecraft security is what enables the offensive character of agent-operational work."

Experience has shown that ensuring the security of intelligence activities demands not only operational competence, but also the conscientious execution of duties under cover, and proper conduct in daily life. Strict adherence to tradecraft requirements is one of the key conditions for successful intelligence operations.

In recent years, the adversary has become more informed about the methods employed by Soviet foreign intelligence, particularly the KGB. The enemy actively exploits previously known data in its counterintelligence operations against Soviet intelligence abroad. In this situation, the necessity of intensifying tradecraft measures—including within the territory of the Soviet Union—has become increasingly evident.

In this study, an initial attempt is made to characterize the particular features of clandestine intelligence activity conducted from within the territory of the Soviet Union. Previous discussions of tradecraft in Chekist literature have primarily addressed it in the context of work conducted by "legal" KGB stations abroad, along with the question of securing the recruitment process.

The objectives of this analytical overview are threefold: first, to summarize the operational tradecraft experience accumulated by Directorate "RT" of the First Chief Directorate (PGU) in the creation and operational utilization of cover arrangements; second, to examine certain features of legend development, the use of encryption techniques, and deception methods in the course of agent-operational activity; and third, to develop recommendations aimed at enhancing the level of tradecraft security in the activities of case officers, operational teams, and field sections of Directorate "RT" PGU, as well as the initial operational units of the KGB-UKGB.

These objectives determine the structure of the present work.

In preparing this overview, the authors consulted directive documents of the KGB of the USSR, orders and instructions issued by the PGU on securing the safety and secrecy of foreign intelligence operations, materials from Directorate "RT," specialized literature from the operational library, and the results of interviews with experienced operational personnel of Directorate "RT" PGU on the subject of tradecraft in agent-operational work conducted from within the USSR.

This analytical overview is intended for use in the practical work of operational personnel of Directorate "RT" PGU, as well as officers serving in the initial operational units of the KGB-UKGB.

PART I

TRADECRAFT SECURITY – A KEY PRINCIPLE AND PREREQUISITE FOR ORGANIZING EFFECTIVE INTELLIGENCE OPERATIONS

Introduction

V. I. Lenin referred to tradecraft security as a special form of clandestine political struggle—its most critical condition and foremost organizational principle. Guided by Leninist precepts on the "art of tradecraft," the KGB has accumulated extensive experience in the covert confrontation with adversaries.

Under current conditions of heightened confrontation and intensified subversive activity by hostile intelligence services, strict adherence to tradecraft procedures has become an uncompromising requirement for every intelligence officer. As shown by analysis of recent provocations by the adversary targeting personnel of the KGB's foreign intelligence service, nearly two-thirds of these provocations were successful due to the adversary's awareness of errors and lapses committed by KGB officers themselves. Foremost among these are violations of tradecraft principles and the failure to properly integrate political activity into intelligence work.

Tradecraft security is an indispensable condition for the effective functioning of the KGB's foreign intelligence apparatus.

The *Dictionary of Chekist Terminology* offers a comprehensive definition:

"Tradecraft Security (Konspiratsiya) is one of the core principles of intelligence work, ensuring its covert character. Tradecraft security entails maintaining secrecy from the adversary, from unauthorized individuals, and from any persons not directly involved in a given intelligence operation. It encompasses the subjects and objects of the activity, its goals, means, methods, forms, and outcomes. Tradecraft security covers all aspects of intelligence work and is supported through

organizational and administrative measures, operational techniques, personnel training, and ideological-political reinforcement measures (such as cover arrangements, encryption of operational participants, secure document handling, techniques of cover identity construction, disinformation, development of tradecraft-conscious behavior, fostering vigilance, and a creative approach to intelligence work, among other skills)."

1. The Operational Scope of Tradecraft Security

Tradecraft security must not be interpreted merely as adherence to a prescribed set of behavioral rules during the execution of operational tasks. The ability to operate covertly in a variety of circumstances—including those not directly related to official duties—is a defining professional trait of the intelligence officer.

The continuous refinement of tradecraft techniques is among the most critical tasks facing operational personnel. The tools and methods of tradecraft must, in any operational environment, ensure concealment of the intelligence service's involvement in covert actions or other operational activities, the invisibility and deniability of officers, agents, and trusted contacts.

2. Vigilance as a Core Competency

In identifying the place and role of this tradecraft within the broader framework of external intelligence principles of the state security organs, it is essential also to examine the concepts of vigilance and security. The interrelationship of these concepts is self-evident, as the security of intelligence operations and measures is grounded in the constant vigilance of intelligence officers and in the strictest compliance with tradecraft procedures in their everyday conduct—whether operating within the central directorate or in cover agencies.

Vigilance in intelligence work is predicated on a high level of political consciousness among Chekist-intelligence officers, on their constant readiness and ability to recognize threats posed by the adversary to the interests of the Soviet state, to the security of intelligence activity, and to preempt potential damage. Vigilance in intelligence operations is ensured through comprehensive and multifaceted study of the adversary—his special services, operational environment, and surrounding persons—through careful analysis of his actions and behaviors; through the enhancement of political and professional preparation; through maintaining a high level of mobilization readiness of the operational personnel to act under any conditions; and through strict adherence to tradecraft, discipline, and organizational order.

3. Security of Intelligence Operations

Just as vigilance is essential, the security of intelligence operations is likewise ensured through the political and professional development and indoctrination of intelligence personnel, through the system of organization and management, through strict observance of the principles of intelligence activity, and through the implementation of appropriate operational, technical, and other measures. It also requires a creative approach by intelligence officers in solving the tasks assigned to them. The security of intelligence operations means the protection of the intelligence service from the adversary—neutralizing his subversive efforts—which allows the service to function reliably and to fulfill its assigned tasks.

PART II

MAINTAINING TRADECRAFT SECURITY IN THE PROCESS OF ESTABLISHING COVERS AND THEIR OPERATIONAL USE

Introduction

The effectiveness of intelligence work conducted from the territory of the Soviet Union is inextricably tied to the system of covers under which intelligence officers operate. A properly constructed and maintained cover arrangement serves as the principal mechanism for shielding operational activity from adversary detection, enabling Soviet intelligence personnel to interact with targets, conduct surveillance, and carry out recruitment, all while sustaining the outward appearance of legitimate professional engagement. As operational experience has shown, any deficiency in the establishment, preparation, or execution of a cover identity can result not only in the compromise of the officer involved, but in broader counterintelligence setbacks for the service as a whole.

The evolving tactics of Western intelligence agencies, marked by the intensified use of provocations, surveillance, and recruitment operations on Soviet soil, have heightened the urgency of strengthening tradecraft measures tied to cover arrangements. It is no longer sufficient to assign an officer a plausible position; rather, each aspect of the cover must be designed, vetted, and maintained with the same precision and care as the intelligence operation it supports. This includes not only the officer's post and legend-biography, but the institutional behavior patterns, professional competencies, and social interactions that sustain the illusion of authenticity over time.

In this context, the cover is not merely a formality or a bureaucratic designation. It is the foundation upon which all operational work is conducted. The cover must be both functionally useful and doctrinally sound, meaning it must offer legitimate access to intelligence targets, enable routine contact with foreigners or Soviets of operational interest, and allow for the concealment of intelligence objectives beneath a fully plausible professional facade. The role of the intelligence officer under cover is therefore not passive or symbolic; it is an active operational posture that must be rehearsed, internalized, and continuously reinforced by tradecraft discipline.

The following sections examine the specific requirements and operational lessons associated with the organization of cover arrangements and the development of legend-biographies. Drawing upon the accumulated experience of Directorate "RT" officers, and guided by the principle that operational cover is a core element of tradecraft—not merely a supporting device—this part of the manual provides detailed guidance for establishing and utilizing covers both within the USSR and abroad. These guidelines are essential for maintaining the deniability, operational flexibility, and strategic reach of the KGB's foreign intelligence apparatus.

1. The Organization of Covers

Among the factors affecting the quality of intelligence operations conducted from within the territory of the Soviet Union, one of the most important is the provision of cover for operational personnel. Cover is understood to mean an official, documented, and functionally supported place of employment for the intelligence officer, consistent with a given cover identity and role, and necessary for the execution and concealment of intelligence tasks.

Cover-related issues fall within the sphere of interagency coordination, as they involve the provision of official posts within Soviet ministries and institutions to intelligence officers. The decision to assign an officer a particular position under cover is made by the relevant authority.

A cover agency is a government body or public organization of the USSR whose official positions are utilized by the KGB's foreign intelligence directorate to provide cover assignments and to facilitate intelligence activities. Agencies providing cover may include the USSR Ministry of Foreign Affairs, the Ministry of Internal Affairs, the State Committee for Science and Technology (GKNT), TASS, the Academy of Sciences, the USSR Committee for Cultural Relations with Foreign Countries (SSOD), academic institutes, and others.[2]

Intelligence officers working under the guise of cover agencies on the territory of the Soviet Union are considered active-duty officers of the KGB's First Chief Directorate (PGU) and are formally attached to it (by agency affiliation).

Recent KGB and PGU directives, instructions, and other normative documents have placed growing emphasis on resolving issues related to cover. Special attention is now paid by the intelligence services to the overall personnel assigned under cover. Underestimating the role of cover can result in the exposure of intelligence officers and their operational contacts.

In resolving matters related to the selection and use of positions within cover agencies, the KGB's foreign intelligence service adheres to the following principles:

First, a given position, when properly integrated with tradecraft protocols, must allow for plausible concealment of intelligence activity—or at least of certain aspects thereof—under the guise of the officer's official duties within the cover agency. Operational practice has shown that some intelligence actions, by their external appearance (and at times even in substance), bear strong resemblance to actions routinely performed by intelligence officers in the line of duty or under their official cover.

Second, the official post must facilitate the acquisition of intelligence from overt sources, as well as the conduct of recruitment operations and active measures.

Selecting a cover arrangement that is most suitable for the execution of operational tasks—and which aligns with the professional background, skill set, and suitability of a particular intelligence officer—is often far from straightforward. A poorly chosen cover frequently results in ineffective operational work or, worse, in the exposure of the officer's true role.

For this reason, a comprehensive assessment of a cover's suitability for fulfilling intelligence objectives, together with oversight from the leadership of relevant departments and directorates and a careful appraisal of an officer's level of preparation, is of exceptional importance. It is not uncommon for valid opportunities to be overlooked due to erroneous assumptions that a particular position within a cover agency is unsuitable. To prevent such mistakes, it is necessary to conduct a detailed study of all available cover possibilities.

As is well known, all intelligence personnel assigned to positions within cover agencies are tasked with a strictly delineated set of long-term duties and obligations. Broadly speaking, these tasks correspond to one of the principal lines of effort in KGB foreign intelligence operations: political intelligence, scientific-technical intelligence, or counterintelligence against foreign services. Consequently, the potential of any given cover arrangement can only be effectively realized when it is matched to a specific operational focus, and when operational tradecraft requirements are strictly adhered to.

For conducting intelligence operations from within the territory of the USSR along political lines, the most suitable cover positions are those of diplomats, journalists, researchers, and employees of governmental and public organizations. These cover roles offer intelligence officers access to foreigners, including those affiliated with or employed in institutions and organizations that the KGB's foreign intelligence directorate considers primary or intermediate targets for penetration.

The correct selection of a cover agency assists in aligning the officer's operational interest with a specific domain of intelligence collection. For example, journalistic cover permits attendance at receptions, press conferences, and other public events. In the West, journalists are regarded as engaged in a transactional information economy, treating information as a commodity to be bought, sold, or exchanged. This allows the intelligence officer to exploit a legitimate avenue for information gathering on a wide array of topics, while simultaneously offering opportunities for recruitment operations and active measures.

Covers based in foreign trade organizations enable intelligence officers to establish contacts with representatives of business circles who have access to sensitive information on economic or scientific-technical matters. These individuals often also maintain significant political or governmental connections within the target countries.

Covers within academic institutions, public organizations, and youth groups provide favorable conditions for accessing a recruitment pool among foreign students studying in the Soviet Union, as well as among youth organizations and public circles of Western countries.

Regardless of the operational track an officer is assigned to, he must always be mindful of how to most fully and effectively exploit the resources of his cover agency in support of intelligence objectives—using cover not merely as concealment, but as a platform for conducting active intelligence work under the guise of legitimate duties.

Every intelligence officer must thoroughly understand the structure and operations of their cover agency in order to credibly sustain their cover identity, and to seamlessly integrate intelligence tasks with the overt functions of the post.

To support a range of operational activities, in addition to long-term cover arrangements, Directorate "RT" extensively employs both single-use and short-term covers.

A single-use cover is employed in instances where it is certain that, following the completion of a specific task, the operative making use of the cover will not encounter the target again. An example would be an operational contact between a KGB officer posing as a militia officer and a subject under investigation, ostensibly for a minor legal infraction. Single-use covers are also sometimes employed during special operations, such as when an operative poses as a journalist, hotel staff member, or representative of a civic organization.

A defining feature of single-use cover is that the operative is not officially employed by the cover institution, nor does he necessarily maintain continuous contact with it or possess documentation verifying the cover. In certain cases, to provide support for a single-use cover, the intelligence service may involve agents or trusted individuals affiliated with the institution, or actual staff from the cover organization from which the operative purports to originate.

In the operational experience of Directorate "RT", short-term covers are also employed. These differ from single-use covers in that they are applied for a relatively brief but bounded period. Such covers are particularly useful in establishing and maintaining contact with foreign nationals of intelligence interest who are temporarily present in the Soviet Union. The legend for short-term covers is prepared in accordance with the operational tasks to be fulfilled during that period. The plausibility of the short-term cover is often reinforced by consistent behavior on the part of the KGB agent and his known associates.

In the pursuit of operational objectives, both single-use and short-term covers are utilized not only by case officers but also by agents of the KGB of the USSR.

Thus, official covers within Soviet institutions provide operatives with sufficiently credible and persuasive legend identities for the conduct of intelligence work and enable the establishment of natural contact with foreigners, allowing interaction with them both in official and unofficial settings. At the same time, it is necessary to consider the well-known suspicion foreigners often harbor toward Soviet citizens who interact with them professionally, as they frequently suspect such individuals of being KGB operatives or agents. Therefore, it is of paramount importance to use cover identities in such a manner that they do not raise any suspicions regarding our interest in the subject.

The official position held by the intelligence officer within the cover organization must meet the following criteria: it must be of interest to foreigners, allow the operative to regularly interact with them, and must be executed with strict adherence to tradecraft measures. Any exposure of the operative could compromise the cover institution itself.

2. The Legend and Operational Biography of the Intelligence Officer

Operational experience indicates that the level of tradecraft in the activities of Directorate "RT" case officers depends not only on how appropriately and purposefully the cover identity has been selected and implemented in accordance with the intelligence objectives, but also on the extent to which the officer's legend — his personal operational biography — has been thoroughly and competently developed. This includes consideration of the nature of the operative's official position, the character of the duties associated with it, and the operative's own personal and professional qualities.

The legend-biography constitutes the core legend of the intelligence officer. It includes a specially tailored, internally consistent set of biographical data (motives, reasons, circumstances, etc.) by means of which the officer conceals from adversaries and his surrounding environment his affiliation with Soviet intelligence, both generally and in the context of specific operations. Despite any fabricated elements it may contain, the legend must be plausible, inspire trust, and withstand scrutiny.

When constructing the legend-biography, due consideration must be given to the officer's cover identity, the operational conditions under which intelligence work will be carried out, and the specific tasks to be fulfilled.

The stage of introducing the intelligence officer into the cover organization and adapting him to its work environment is considered the beginning of a comprehensive operational deployment of the cover identity and legend-biography. An essential condition for embedding an officer into a cover organization is the thorough preparation of the legend-biography and the creation of all necessary personal documentation. To accomplish this successfully, the officer must become thoroughly acquainted with the general requirements and procedures for employment within the cover institution.

According to existing requirements, a prospective intelligence officer entering a cover organization must possess documentation aligned with his legend-biography. This includes a passport, work record book, Communist Party (or Komsomol) membership card, military and trade union cards, union membership ledger, and, if needed, a dues payment record, salary certificate, employment or academic reference, and educational diplomas (in some cases, also a certificate of residence).

Core documents, such as the passport, marriage certificate, educational diploma, may be genuine. However, the party card (Komsomol card) is typically fabricated, as are the employment record book, military and trade union cards, and the professional reference from the previous (legend-based) workplace. All are prepared in strict accordance with the legend-biography.

The beginning of work in a cover organization constitutes a challenging period in the life of an intelligence officer. In any institution, a new employee invariably attracts increased attention. For some time, he remains in the spotlight of the collective. Colleagues naturally wish to learn more about the newcomer, his personal and professional traits, demeanor, lifestyle, family status, hobbies, and the like. During this introductory period, shared interests and values, common acquaintances, and mutual hobbies may generate a sense of reciprocal sympathy and lead to the formation of friendly relations. Subsequently, a period of more in-depth observation of the new officer's professional and personal qualities follows. Gradually, the collective becomes accustomed to the newly assigned individual, forms a certain opinion of him, and he ceases to be the focus of special attention. It is during this period that the legend-biography of the intelligence officer undergoes its most critical test, his ability to live it convincingly. Even a minor inconsistency, falsehood, or uncharacteristic action may lead to exposure.

Tradecraft requirements dictate that the intelligence officer strictly adhere to his legend-biography and conduct himself in accordance with his official cover throughout the entire duration of his assignment within the cover organization. Operational experience demonstrates that it is precisely deviations from established behavioral norms within the organization that often result in the exposure of the officer and can have serious operational and, at times, political consequences.

This raises the question: how should the officer ensure the operational security of his intelligence work?

First and foremost, he must be capable of convincing others—by his behavior—of the authenticity of his legend. An officer should not externally differ in any way from other personnel of the organization. In the eyes of foreign colleagues, he must present himself as a competent and credible professional. However, not all officers succeed in this regard. Some, for example, draw attention to themselves through insufficient training related to their cover role, careless performance of their official duties, inappropriate demeanor at work, or failure to conform to the established internal routines of the organization.

The officer's conduct must align with the known behavioral norms and individual peculiarities of those around him, his personality, habits, demeanor, and so on, meaning it must appear natural. Naturalness is the principal condition of effective covert behavior. It must always be context-specific. It requires the officer to operate in a manner appropriate to the environment, whether among Soviet citizens or foreigners, during official or intelligence-related assignments (establishing initial contact with foreigners, developing operational relationships, collecting intelligence, and so on). For example, demonstrative displays of curiosity or detailed questioning about military-industrial matters might seem acceptable in casual conversation among technical personnel. Such conduct, however, on the part of an intelligence officer posing as a senior engineer of a foreign trade association would arouse suspicion due to the misalignment with his professional interests.

Particular attention must be paid to the officer's conduct in scenarios specifically contrived by the adversary for the purpose of exposing his intelligence activities, especially during temporary assignments abroad under official cover. Before departure on a short-term mission abroad via institutional channels, the operative must thoroughly master the essential knowledge related to his cover role and complete the necessary specialized training. An individual tasking plan, developed jointly by the officer and the head of the relevant intelligence department and approved by the management of the appropriate department of Directorate "RT", must contain provisions for the development of a cover legend and supplemental measures for backstopping (e.g., a legend explaining the officer's departure from the cover organization's workplace for operational reasons).

The operative's behavioral pattern must be guided by the nature of his cover assignment and the stated intelligence mission, the operational environment, the tools and methods employed, and the national-psychological traits of the individuals with whom he interacts, along with the officer's own personal characteristics. All these factors must be accurately assessed to select the most appropriate tradecraft approach for accomplishing the intelligence task.

Officers of the active reserve of the First Chief Directorate (PGU) assigned to cover organizations are formally part of the intelligence department. It is essential that, in the execution of intelligence tasks, day-to-day contacts between case officers and their operational colleagues appear natural, are easily explained to outsiders, and are convincingly substantiated by the fulfillment of the official duties of the officer's cover position.

PART III

FEATURES OF THE LEGEND DEVELOPMENT, DISGUISE AND DECEPTION TECHNIQUES IN THE PROCESS OF AGENT-OPERATIONAL ACTIVITY

Introduction

Tradecraft in the context of agent recruitment is expressed through the implementation of security measures during planning and execution, ensuring the safety of each recruitment operation, and preserving in strict secrecy from the adversary and third parties all information related to recruitment targets, as well as the personnel, resources, and methods of the KGB's foreign intelligence service employed in recruitment activities. The adversary's counterintelligence services actively seek to hinder recruitment work—primarily by inserting provocateurs into the KGB's agent network. Consequently, any failure to adhere to tradecraft protocols or errors committed during the recruitment phase inevitably lead to the collapse of agent operations.

Enemy counterintelligence services continuously refine their methods and practices, particularly in counter-recruitment: collecting, comparing, analyzing, and synthesizing data to identify indicators of intelligence activity. This includes monitoring Soviet citizens whose professions require interaction with foreigners, both within the USSR and abroad. The counterintelligence services of many capitalist states have been able to identify certain hallmarks of Soviet citizens involved in intelligence work. These include behavioral traits or occupational patterns they interpret as indicative of KGB affiliation.

Contributing factors to such suspicions may include poor selection of cover identity, improper use of the assigned cover (i.e., professional incompetence), errors in tradecraft, or careless execution of duties by the officer under cover. Suspicious conduct may also stem from behavior inconsistent with the assumed role, as well as activities typical of intelligence work but lacking appropriate concealment—thus betraying the officer's true affiliation despite sophisticated masking techniques. Even seemingly minor personal indiscretions, lack of caution, or temporary lapses in vigilance may compromise the operation.

Exposure of a recruitment effort can occur not only due to enemy counterintelligence efforts, but also as a result of unguarded behavior by the target of recruitment or errors on their part—sometimes caused by incompetent guidance from handlers or lapses on the part of the officers themselves. In order to create conditions conducive to covert behavior by the target of recruitment, the intelligence officer must first ensure concealment of their contact with the target from surrounding individuals, and second, must obscure from the target—especially during the initial stages—the officer's true affiliation with the intelligence service and the genuine nature of their intentions and objectives.

A recruitment operation can only be successfully masked if the officer consistently adheres to tradecraft principles throughout every phase, and if the target takes measures to preserve the secrecy of their contact with the Soviet representative.

1. Tradecraft Measures in the Stages of Target Identification and Recruitment Operations

Initial Phase of Target Identification and Recruitment.
At this stage, it is critical to select the appropriate method for establishing initial personal contact with the foreigner. From the outset, the officer's contacts must be plausibly covered, taking place either in the context of the officer's assigned post under the auspices of a Soviet institution, or through seemingly "accidental" encounters at events such as forums, exhibitions, conferences, and the like, or else on a "neutral" basis. In establishing direct contact with the target, the use of an intermediary, whether an agent or a trusted liaison, can also serve the legend. In all cases, a thoroughly developed cover identity and careful preparation for the encounter are essential.

Experience has shown that when the approach to a target occurs from a position associated with the officer's official cover assignment, tradecraft concealment is at its most effective. The art of tradecraft at this stage depends heavily on the officer's ability to mask both their true intentions and their interest in the target while dealing with foreigners.

It is a different matter if the contact is established outside of the official cover structure (e.g., at a reception, conference, etc.). In such cases, the circumstances vary. For instance, the presence of hostile intelligence personnel or their assets is possible at such events. Therefore, from the outset, the officer must take appropriate measures to obscure the fact of the acquaintance, specifically, to avoid showing the target more attention than to others and not to signal any particular interest in them.

With regard to establishing initial contact based on a referral received from a "legal" KGB Station, either through agents or case officers, the legend under which the target is approached takes on particular significance, especially if the target is already located within the USSR. In cases where the target is invited into the country on the basis of a referral, it is important to select a host organization whose profile aligns with the target's activities and which possesses the capacity to facilitate their study.

Thus, the correct choice of method and cover structure for initiating first contact, from within an appropriate agency, combined with a well-constructed operational legend, enables secure handling of the initial phase of the recruitment process. The cover identity used for the initial contact with the foreigner must satisfy at least two criteria: it must be natural, that is, it must fit seamlessly into the surrounding circumstances, and it must be flexible, meaning it should permit the continuation or termination of the relationship as necessary, without arousing suspicion from hostile services.

The naturalness of the initial contact legend presumes not only a logical reason for the officer's presence at the given location or event, but also its consistency with the functions and responsibilities assumed under the cover.

Operational experience shows that initial contact between a KGB officer and a foreigner, depending on specific conditions, may be plausibly explained by official duties, shared professional or personal interests, the foreigner's desire for support or cooperation, and similar factors.

A well-devised cover legend for the initial acquaintance assists the officer in encrypting subsequent interactions with the foreigner and shielding their true operational significance.

However, operational records indicate that some intelligence officers begin considering the encryption of their contact with a foreigner only once they sense that the recruitment development may lead to a favorable outcome.

This failure to meet the requirements of tradecraft concealment during the initial phase of recruitment operations stems from an underestimation of the importance of the contact-establishment process itself. This is evidenced by case officer reports concerning initial contacts with foreigners, which at times omit not only the circumstances of the meeting but also details such as the presence of others, the location, and even the date of the initial encounter.

One of the conditions for ensuring the security of first contact is the thorough preparation of the intelligence officer. This includes gathering minimal but essential information about the target, choosing the time and place, and creating a setting conducive to contact. It also entails preparing the legend under which the contact will be made and later developed, taking into account the target's national-psychological and individual characteristics. It is essential to select appropriate themes for conversation and plan further encryption measures for future meetings. Furthermore, it is necessary to devise methods for detecting possible hostile provocations by adversarial intelligence services.

It is of utmost importance to the intelligence service to protect the recruitment target from exposure to counterintelligence from the very beginning of the operation. Accordingly, the full arsenal of methods, techniques, and resources available to intelligence is deployed to ensure the security of recruitment operations: these include agents among foreigners and Soviet citizens, trusted intermediaries, operational communications, resources of the 2nd Chief Directorate, 5th Chief Directorate, "legal" KGB Station assets, operational-technical means, surveillance, and more. Naturally, the choice of which of these tools to apply depends on the concrete circumstances unfolding during the development of the operation.

A key factor in ensuring the concealment of recruitment activities is the use of agents selected from among foreigners and Soviet citizens who have been sufficiently vetted and established as reliable. These agents may be utilized to establish and maintain plausible cover for the intelligence officer or the target. For example, when developing operations involving foreign students in the USSR, it may be advisable to use foreign agents as intermediaries to achieve the most effective concealment outcome. For such agents, a convincing cover story and other measures are adopted to prevent any mutual exposure of the recruitment target and the agent. Operating in a natural and familiar environment, without arousing suspicion from those around them or falling within the field of vision of the adversary's counterintelligence, the agent-handler (whether a foreign or Soviet agent) ensures the acquisition of necessary information. This includes the ability to determine the developmental basis of the operation, monitor its progress, introduce timely adjustments, avoid compromise, and precisely determine the optimal point to terminate the development process.

When trust-based communications are employed, the scope of assigned tasks inevitably narrows, operational planning becomes simpler, and the importance of maintaining a robust legend increases. The primary function of such contacts is to clarify the "dark spots" in specific areas of interest. Throughout this process, it is imperative to ensure that the activities of the trusted agent do not extend beyond the bounds of the legend or arouse the interest of the subject, thereby avoiding any actions that might reveal the operation or expose the KGB officer. The entire development must be constantly monitored to prevent interactions with the foreign target that could jeopardize the operation through inadvertent disclosure.

Valuable information can also be gathered through passive observation of the target under development. From the target's appearance, behavior, and reactions, many insights relevant to the recruitment effort can be gleaned.

Available records confirm that the intelligence services of not only capitalist but also several developing countries conduct counterintelligence efforts against their nationals working abroad, particularly those currently in the USSR. Moreover, foreign intelligence services frequently use entry into the USSR as a channel for inserting operatives tasked with various intelligence assignments.

To prevent hostile penetration of the KGB's agent network, it is essential to carry out a comprehensive suite of vetting procedures at the initial stage of development.

Years of operational practice within the USSR have produced a diverse range of techniques and methods for verifying the trustworthiness of targets under recruitment. The choice and application of these techniques depends upon the specific operational objectives, the personal characteristics of the foreigner, and the context of their intended use. During the vetting process, it is necessary to determine the degree to which the foreigner may be aware of counterintelligence procedures, the nature of KGB operations, and any potential ties to intelligence services, interest in classified matters, contacts with the embassy of his home country, and whether he may be an adventurist or pursuing hostile aims, etc. To verify the sincerity of the subject and ensure that he does not disclose the fact or nature of his relationship with a Soviet intelligence officer, it is standard practice to approach the subject using vetted agents drawn from among foreigners or Soviet citizens. Such operational combinations usually produce good results and are considered a highly effective method of conducting counterintelligence verification within the USSR.

When assessing foreigners, one must refrain from drawing premature conclusions based solely on initial information. Hostile intent or conduct by a given foreign individual may only become apparent at a later stage of their presence in the USSR.

One directive from the Chief of the First Chief Directorate states:

> *"At present, when the adversary has chosen the infiltration of its own agents into the Soviet intelligence apparatus as a primary method of obstructing and discrediting the work of Soviet intelligence, particular importance must*

be placed on conducting vetting procedures for every foreigner who expresses willingness to cooperate with Soviet intelligence."

It is further noted that:

"Any shortcomings in the execution of vetting procedures, or their unqualified implementation, can not only result in the infiltration of the agent network, but may also inflict serious political damage to the Soviet Union."

A particularly vulnerable point from a tradecraft perspective in working with agents is the recruitment interview. This must be conducted strictly in accordance with a specially formulated operational plan. During the preparatory phase, it is essential to incorporate effective measures to ensure the security of the intelligence officer and the agent under recruitment. A poorly handled recruitment can compromise the agent-operational apparatus and, in some cases, cause political damage to our country.

When choosing the place and time for the recruitment interview, it is necessary to consider that the meeting location must meet the requirements of tradecraft and ensure the safety of the recruiting officer in the event of a recruitment failure. The meeting may be conducted in a restaurant, hotel, safehouse, etc. These venues must exclude the possibility of the target encountering compatriots. To localize the consequences of a failed recruitment attempt, operations involving foreigners are preferably conducted not in Moscow but in other cities of the Soviet Union.

The recruitment interview has a strong psychological impact on the target. By agreeing to cooperate in intelligence activities, the recruit takes a known risk and entrusts their safety to the intelligence service. Therefore, the recruiting officer must convince the recruit that the intelligence service is equally committed to ensuring his security.

The recruitment interview may be conducted either in the name of the KGB or through the cover identity of the department. Typically, high-risk recruitment operations are conducted by officers of the KGB Central apparatus. Some agents recruited on the territory of the USSR remain under the control of officers from the "RT" Directorate of the First Chief Directorate. This occurs for several reasons: the agent's remoteness from KGB Station locations, lack of viable meeting conditions in the target country, or the agent's high social status, which precludes maintaining cover contacts with KGB Station personnel. In such cases, contact is maintained exclusively with designated Soviet handlers. Another factor is that the agent has not yet been fully vetted and validated.

If a foreign national remains in the USSR for some time after recruitment, it is necessary to consolidate the cooperation by obtaining classified information, written statements concerning members of his environment, signed financial receipts, etc. To be fully assured of the agent's sincerity and loyalty, it is necessary to assign him specific operational tasks. It is not advisable to formally transfer the agent to KGB Station control until this process has been completed, and he becomes involved in practical cooperation, validated and tested in the fulfillment of intelligence tasks.

Regarding the question of whether it is possible to define the duration of a recruitment operation, or more precisely, within what time frame it should be completed in the interest of security, there is no clear-cut answer. Nevertheless, it is necessary to establish approximate time limits for the completion of various stages of a recruitment operation in order to fully exploit the favorable conditions present within the Soviet Union and utilize the means at hand to achieve the final objective, avoiding unjustified delays and maintaining strict tradecraft.

2. Security Tradecraft in Operations Involving Agents and Confidential Contacts

In agent operations, the foremost requirement is adherence to the principle of tradecraft. This means that the agent's affiliation with Soviet intelligence must be known only to the officers or agents who are required to be informed in order to fulfill their own responsibilities. Information concerning the methods and means used by intelligence officers and other classified data must be disclosed to the agent only to the extent required to carry out the assignment. Every in-person meeting must be carefully prepared. All necessary precautions must be taken to ensure the agent's security. Secure forms of communication must be actively employed, and the handling of the agent must be conducted in such a way that any information obtained does not result in the exposure of its source.

Furthermore, contact with the agent must be maintained strictly within the bounds of tradecraft. Whenever possible, the agent should not be drawn into operations involving contact with the agent network or with targets under development by other officers, particularly those not connected to the agent's handler. Operational reports and dossiers on the agent or the case should contain only the minimum necessary information regarding persons of interest, to reduce the risk of compromise.

Strict adherence to the requirements of tradecraft ensures not only the security of working with an agent but also maximizes the agent's operational potential for successfully completing intelligence tasks. Therefore, the case officer must take all necessary measures to conceal his relationship with the agent, particularly from the agent's compatriots, and to reliably provide cover for their contact.

Naturally, if the agent maintains regular professional contact with the cover institution, the officer, making use of the natural environment, should conduct meetings with the agent almost openly, in view of colleagues, concealing only the true nature of the interaction. If the agent is in contact with an operational officer of the Center, special attention must be paid to the meeting location. As a rule, such meetings occur in safe houses or other preselected locations where contact between the agent and his compatriots is unlikely.

Operations on Soviet territory typically involve meetings with agents abroad, in third countries. In such cases, tradecraft measures are employed that mirror those used by KGB "legal" Stations.

The plan for a case officer's meeting with an agent abroad is usually coordinated with the KGB's geographical department and the "legal" Station. These entities assist in selecting secure meeting sites, developing communication protocols, addressing counter-surveillance measures, creating the cover story for the officer's departure, and establishing the operational legend for the meeting, both for the officer and the agent.

At all stages of agent work, particular emphasis must be placed on verification and consolidation through active operations. This includes the use of agent networks, operational-technical measures, and physical surveillance.

A critical phase in working with agents from among foreign nationals is preparing them for operations after they leave the Soviet Union. At this stage, alongside operational preparation, significant attention must be given to ensuring that the agent upholds tradecraft during his time abroad, retains the communication protocols in memory, etc. Equally important is training the agent in methods of covert inquiry and information collection, both independently and through his own contacts, as well as in the processing and safekeeping of such information.

When an agent is transferred to the control of a KGB station, the communication procedures must be established in advance: overt, primary, and fallback meetings, emergency summonses. It is advisable to instruct the agent in various forms of impersonal communication, including dead drops, immediate transfers, etc. If necessary, the agent must also be trained in the use of operational-technical communication tools. Theoretical knowledge must be reinforced through practical exercises.

It is known that Western intelligence services maintain watchlists of individuals residing in the USSR, categorizing them as potential "agents" of the KGB. In light of this, in certain cases, it is necessary to take special measures to conceal the very fact of an agent's stay in the USSR.

The operational preparation of a foreign agent, as a rule, must be carried out in close coordination with the geographic department of the First Chief Directorate (PGU), with direct participation from both the receiving and the preparing sides— ensuring high-quality preparation of the agent who is intended to be transferred under the control of a station. Therefore, agent preparation must follow a plan jointly approved by the geographic department of the PGU and Directorate "RT" (Counterintelligence Operations), taking into account the specific operational environment in the country to which the agent will be deployed.

Tradecraft measures for working with trusted contacts inside the USSR are applied with limitations. The officer appears legally under cover of the authorized institution. A different situation arises abroad, when the officer or agent travels for a meeting with a trusted contact. In order to avoid compromising the operational relationship with the contact to adversary services, it is essential to observe all tradecraft measures appropriate to the position of a "legal" KGB station chief. Therefore, prior to departure, it is critical to develop a cover story for the meeting—so that the meeting appears to be an official contact. For example, to meet with a trusted contact in the United States, it is presented as a scientific delegation, an academic visit, or participation in an official delegation, etc.

When working with trusted contacts within the USSR, the intelligence officer generally disguises meetings under the pretext of addressing matters of mutual interest through the cover institution. Meetings are conducted openly. Many encounters with foreigners accredited in the Soviet Union (diplomats, businesspeople, journalists, etc.) are regular in nature. For instance, contacts with a foreign diplomat may appear normal, as it is within his professional duties to frequently visit the USSR Ministry of Foreign Affairs.

Engagement with certain trusted contacts can become complicated when, due to political, official, business, or scientific considerations, they may seek associations with personnel from Soviet institutions or Soviet organizations known within specific circles. In such instances, particular importance must be given, on the one hand, to a reliable and convincing legend to cover such contacts, and on the other, to maintaining the secrecy of the relationship between the trusted contact and the intelligence officer. The legend must be able to withstand scrutiny by the adversary's counterintelligence services and ensure the continuation of the trusted relationship, even when the adversary is aware of regular meetings. The cover story must align strictly with the official status and genuine interests of the trusted contact, as well as the public role and activities of the intelligence officer. In operations involving trusted contacts on Soviet territory, special importance is placed on creating plausible legends for their visits to the USSR, including justification for the necessity of such visits and their sources or means of funding.

For example, an intelligence officer maintains a trusted relationship with "F", a representative of a commercial firm operating in the Near East. Due to close ties with political and academic circles in several countries, "F" has access to valuable intelligence. His relationship with the Soviet representative is explained under a business cover: securing contracts with Soviet institutions. The officer assumes the role of company liaison, involved in negotiating commercial agreements between "F" and Soviet foreign trade organizations. The intelligence officer worked successfully with "F" because the cover story was convincing and the counterintelligence services could not initiate any provocations against "F".

Trusted contacts are instructed cautiously and, as a rule, in a veiled manner—primarily regarding the methods for concealing the nature of the relationship with the intelligence officer. As for specific methods used to execute intelligence tasks, these are never disclosed to the trusted contact, and the foreigner does not receive any specialized training on such matters.

For example, a scholar, diplomat, or journalist is always seeking specific information that serves their own interests. The intelligence officer may assist them through advice or guide their interest toward intelligence-related matters. Businessmen who provide materials or samples on a confidential basis may also be leveraged, drawing on their own experience in commercial activity. No special operational training is required for foreigners with whom trusted relations are maintained and who participate in active measures, such as publishing articles, delivering public speeches, submitting parliamentary inquiries, etc. However, this does not eliminate the need for bilateral discussions of their activities, proposed methods, and decision pathways for specific intelligence tasks. Planned measures must be developed in depth, with close attention to the security aspects of the operation.

One feature of working with trusted contacts is that, in establishing and maintaining such relationships, the intelligence officer typically does not disclose affiliation with the security organs, nor engages in actions that could be clearly identified as characteristic of intelligence operations. To the foreigner, the relationship should appear as collaboration of a non-intelligence nature—whether political, commercial, scientific, or otherwise. Although the foreigner may eventually suspect the connection exceeds routine departmental interests, and that his counterpart may belong to another service, this may still be acceptable to him. He himself may also exploit the trusted relationship to address such matters that could not be handled within the confines of purely official relations.

Naturally, the requirement to prevent the exposure of intelligence officers before trusted contacts cannot be regarded as absolute. In complex operational conditions, certain exceptions are unavoidable—especially given that the level of operational development may bring trusted relationships close to agent-level status.

Trusted relationships often serve as an intermediate stage on the path toward establishing full agent relations during the gradual process of involving a foreign national in cooperation with Soviet intelligence. This form of collaboration is frequently the only viable option in the operational handling of high-profile political figures, scientists, businessmen, and journalists visiting the Soviet Union.

CONCLUSION

Analyzing the operational practices of officers from the active reserve of the First Chief Directorate (PGU) across departmental lines, one may conclude that adherence to the requirements and rules of tradecraft security in intelligence operations is of critical importance for the successful fulfillment of the operational tasks assigned to them. This analysis allows us to offer several recommendations that may contribute to raising the level of operational security in intelligence work conducted from Soviet territory.

The effectiveness of intelligence operations is largely determined by the personal qualities of the intelligence officer. This necessitates very high standards for the system of selection, training, indoctrination, and assignment of operational personnel within cover organizations. Therefore, when assigning an intelligence officer to a cover position and developing their legend (cover biography), consideration must be given to their basic training (professional and linguistic), civilian specialty, and the feasibility of maintaining the cover to fulfill intelligence tasks. In situations where the cover provided by the department is inadequate for the full scope of operational tasks, it is advisable to redirect officers to work under the guise of other institutions (e.g., public organizations), where conditions for operating in foreign countries are more favorable.

As a rule, compromise occurs as a result of insufficient professional vigilance on the part of personnel. In this regard, it is considered expedient to implement a systematic study by personnel of existing orders and directives concerning tradecraft, regardless of their place of work or assignment. These should define the forms of effective control necessary to ensure compliance with operational security rules in practice.

The organization of such work would contribute to the more effective implementation of the Order of the Chairman of the KGB of the USSR No. 01000 of 19 February 1986, which mandates "ensuring the thorough verification of knowledge of the requirements and rules of tradecraft in addressing issues related to the assignment of official duties."

In order to raise the level of operational security among officers of the active reserve of the First Chief Directorate (PGU) operating within departmental cover assignments, it is necessary to:

– Strengthen supervisory control by department leadership and supervisors over the level of professional training of operational personnel and their adherence to the requirements of tradecraft;

– Align the operational regime of intelligence officers embedded in cover organizations as closely as possible with the working regime of those institutions;

– Require heads of intelligence units and supervisors to continuously monitor the recruitment development of foreign nationals to prevent situations that may lead to compromise;

– Constantly ensure that tasks assigned to confidential contacts do not exceed the boundaries of the established legend and do not lead to the exposure of intelligence intentions;

– Streamline the involvement of case officers and targets in agent operations, and ensure proper processing of information received from these contacts, avoiding direct involvement of unrelated personnel;

– Prohibit the compilation of agent or operational target files without clear operational necessity. Ensure such documents contain only the minimum data necessary regarding persons of interest;

– When making decisions on introducing an agent into operational development, carefully weigh their reliability, personal and professional qualities, and their level of operational preparation.

To safeguard the security of recruitment development activities, it is necessary to make broader use of the cover identity of the department, including agent developers and recruiters.

To reinforce legends and acquire practical experience, it is advisable to assign young personnel, already embedded in cover organizations, to short-term foreign assignments.

Prior to the departure of an agent from the Soviet Union for operations abroad, it is necessary to:

– Acquaint them, to the permissible extent, with materials regarding the operational situation in the target country, including information about the activities of local intelligence and counterintelligence services;

– Instruct them in methods of non-contact surveillance of individuals of operational interest;

– Train them in establishing and developing contacts with individuals not in the agent's immediate circle, including the ability to utilize open-source publications to identify intelligence targets.

For the instruction and preparation of especially valuable agents, it is advisable to involve operational officers from the central apparatus of the PGU's Directorate "RT".